DATE DUE

H A B I T A T S

GRASSLANDS

JULIA WATERLOW

RSVP
RAINTREE
STECK-VAUGHN
P U B L I S H E R S
The Steck-Vaughn Company

Austin, Texas

H A B I T A T S

Coasts   **Mountains**
Deserts   **Oceans and Seas**
Forests   **Polar Regions**
Grasslands   **Rivers and Lakes**
Islands   **Wetlands**

Cover: Savanna grassland in Masai Mara game park in Kenya

Title page: Prairie grasslands at Buffalo Gap in South Dakota

Contents page: Rounding up sheep in the Mongolian steppe

To my husband David

Published by Raintree Steck-Vaughn Publishers,
an imprint of Steck-Vaughn Company

Library of Congress Cataloging-in-Publication Data
Waterlow, Julia.
Grasslands / Julia Waterlow.
     p.   cm.—(Habitats)
     Includes bibliographical references and index.
     Summary: Explains how grasslands are formed, the wildlife that inhabits them, and the impact man has on the grassland environment.
     ISBN 0-8172-4518-9
     1. Grasslands—Juvenile literature.
     2. Grassland ecology—Juvenile literature.
     [1. Grasslands. 2. Grassland ecology. 3. Ecology]
     I. Title. II. Series.
     QH87.7.W37   1997
     574.5'2643—dc20   96-46356

Printed in Italy
1 2 3 4 5 6 7 8 9 0 00 99 98 97 96

CONTENTS

1. WIDE EXPANSES OF GRASS

At the edge of the world's forests trees gradually thin out, becoming smaller and fewer. The landscape becomes open, with views far into the distance. Trees give way to small shrubs and grasses; sometimes there are only grasses, covering wide plains like a thick carpet. This kind of habitat is known as grassland.

Natural grasslands lie between regions of wet forests and dry deserts. Rainfall in grasslands is too low for many large plants, such as trees, to grow, but there is enough rain for small plants like grasses. Where grasslands meet deserts, grasses gradually thin out until just small clumps struggle to survive.

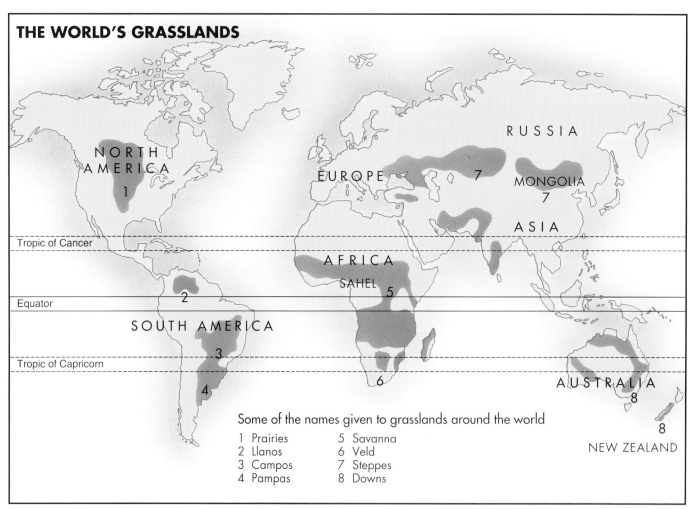

THE WORLD'S GRASSLANDS

NORTH AMERICA 1

RUSSIA

EUROPE 7

MONGOLIA 7

ASIA

Tropic of Cancer

AFRICA

SAHEL 5

Equator

2

SOUTH AMERICA

3

Tropic of Capricorn

4

6

AUSTRALIA 8

8

NEW ZEALAND

Some of the names given to grasslands around the world

1 Prairies   5 Savanna
2 Llanos   6 Veld
3 Campos   7 Steppes
4 Pampas   8 Downs

There are two main types of natural grassland. One type is tropical or savanna grassland. Tropical grasslands lie on either side, to the north and south, of the band of dense forest that grows in hot and humid equatorial regions. Tropical grasslands do not all have the same kind of vegetation; many have a scattering of trees and bushes, and the height of grasses varies depending on rainfall and soil. Although temperatures are usually high, rainfall is seasonal, and this variation throughout the year has a great effect on the growth of plants and the habits of the animals that live there.

The other type of grassland lies in temperate areas of the world and is sometimes known as prairie or steppe. Temperate grasslands are usually found in the middle of continents or in the shadow of high mountains (the rain shadow) where rainfall is low. Trees are rare and mainly grow in hollows or river valleys. The temperature in temperate grasslands varies from very cold in winter to very hot in summer.

More than one third of the world's land area was once covered in natural grasslands. They can still be found on all the continents except Antarctica. Some of the largest areas of tropical grassland are to be found in Africa and are known as savanna. North of the Amazon rain forest in South America, there are the llanos, tropical grasslands in the basin of the Orinoco River, and to the south are scrub and grass regions of Brazil, sometimes called the campos. Large parts of Australia are also covered in savanna vegetation.

Above Elephants graze the tropical savanna grasslands of Africa. Trees and bushes, as well as grasses, often grow in this type of grassland.

Right The wide open spaces of temperate grasslands in Mongolia in central Asia. Many of the people who live there keep flocks of sheep and goats.

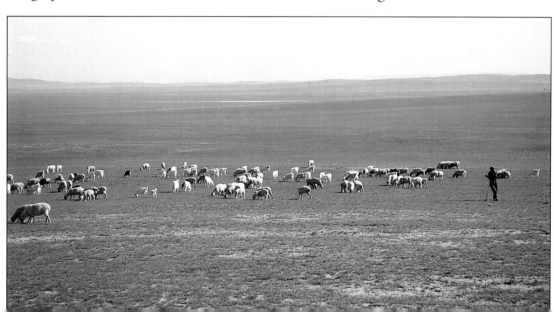

One of the largest areas of temperate grassland in the world is the steppe (from a Russian word meaning treeless plain), which stretches nearly five thousand miles from Europe to central Asia and beyond as far as Mongolia and China. Much of the grassland has been turned into farmland, especially in the western steppes where the rainfall is higher. Only the margins remain in their natural state. In North America, as well, there were once vast grasslands, the prairies, that lay in the region between the Appalachian and Rocky Mountains. Most of the prairies are now plowed and farmed, and the natural grasses have gone.

In the southern continents of the world, temperate grasslands include the pampas of Argentina in South America and the veld (meaning field) in South Africa. In New Zealand, areas of grassland are sometimes called downs. In Africa and Australia, there is not always a distinct line between temperate grasslands and savanna grasslands, and so many plants and animals are common to both areas.

The prairies of North America were once huge areas of natural grassland. Most of the land was plowed and is now used for growing wheat and other crops.

Grasslands can also be found at high altitudes, where the climate is too cold for trees to grow. The largest areas of high altitude grassland are to be found on high plateaus among mountain ranges such as the Himalayas in Asia and the Andes in South America. High altitude grasslands, which include the pamirs of central Asia and the puna of the Andes, are mostly treeless, and only rough grass grows there.

There is one other kind of grassland, one created by humans. In parts of Europe and North America, trees have been cleared and special grasses sown to create wetland meadows, providing rich grazing land for animals. These are not natural grassland habitats but artificial ones, and because just one type of grass is grown and other plants are kept down by farmers, this grassland does not support anything like the same amount and variety of wildlife found in natural grasslands.

Virtually none of the world's natural grasslands now lies undisturbed. In the past, every grassland was once grazed by wild animals; most are now home to domesticated animals or else they have been cultivated to grow crops. Grasslands are usually thinly populated because it is not easy to live in a place with low and often unreliable rainfall and where there are few trees that can be used for fuel or for building shelter. Even so, the increasing population of the world means that there is pressure to use grasslands more and more intensively, and they are fast disappearing as a natural habitat.

In Europe, most of the fields and meadows grazed by cattle are in areas of grassland created by humans. For centuries, trees have been cleared and grasses sown to produce rich pasture.

2. THE CLIMATE AND CAUSES OF GRASSLANDS

The main reason why grasslands occur is the climate. Many grasslands have low rainfall or rains that fall at only one time of year, which makes the climate too dry for forests but also too wet for deserts. The pattern of rainfall and temperatures during the year depends on whether the grassland lies in the temperate or tropical regions of the world.

Temperate grasslands

Most temperate grasslands lie between cool and wet latitudes, where forests thrive, and the hot dry deserts that exist around the tropics of Cancer and Capricorn. They are found in the center of continents or in the rain shadow of great mountain ranges, beyond the reach of rain-carrying winds from the sea. Rainfall is usually between 10 and 30 inches a year.

Both the prairies of North America and the steppes of central Asia lie in the middle of continents and are further protected from rain-carrying winds by mountain ranges. In Argentina, the low rainfall of the pampas is caused by the barrier of the high Andes Mountains, which prevent wet southwesterly winds from reaching the east of the continent.

Winters in temperate grasslands are usually cold and dry. As cold air settles over the land it sinks, creating a zone of high pressure. This high-pressure zone blocks any wet winds that might blow in. When occasionally a little moisture gets through, it falls as snow. The highest rainfall occurs in the spring and summer months. At this time of year, when the sun is nearly overhead and temperatures are high, the hot air rises and creates an area of intense low pressure. Winds can be sucked in from far afield and some bring rain.

Herds of alpaca (a kind of llama) graze the short grasses on the high slopes of the Andes Mountains in South America.

In the prairies of North America, summer rainfall is unreliable (see diagram). Moist rain-bearing winds from the Gulf of Mexico have to battle with dry winds coming over the Rocky Mountains to reach the prairies, and it is never certain which wind will prove stronger. When the summer rains do arrive, they are often accompanied by violent thunderstorms that flatten crops. Uncertain rainfall and heavy storms are a feature of other temperate grasslands as well.

Temperatures can be extreme in regions that are far from the sea's influence. The sea tends to warm up and cool down more slowly than land, and it keeps temperatures moderate near coastal areas. In the prairies and the steppes, summer temperatures can soar to over 90°F and yet in winter they can drop to -20°F or lower. Because temperate grasslands of the Southern Hemisphere, such as the pampas, lie closer to the oceans, temperatures are never quite as extreme.

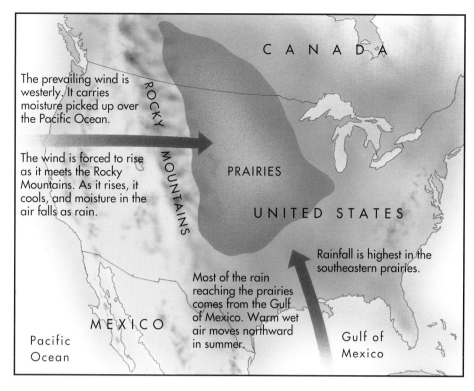

CANADA

The prevailing wind is westerly. It carries moisture picked up over the Pacific Ocean.

ROCKY MOUNTAINS

The wind is forced to rise as it meets the Rocky Mountains. As it rises, it cools, and moisture in the air falls as rain.

PRAIRIES

UNITED STATES

Rainfall is highest in the southeastern prairies.

Most of the rain reaching the prairies comes from the Gulf of Mexico. Warm wet air moves northward in summer.

MEXICO

Pacific Ocean

Gulf of Mexico

Above One of the causes of temperate grasslands is the low rainfall due to the rain shadow effect of mountains. In North America, the Rockies create a barrier against rain-carrying winds.

Left The Rocky Mountains tower above grassland in Colorado.

Left Snow covers the ground during the bitter winters in the Mongolian grasslands of central Asia.

Of all the weather conditions found in temperate grasslands, it is the wind that is most distinctive. With no trees to slow them down, howling winds streak across the open land. Each grassland has a particular wind that blows for some of the year: the chinook of the prairies, the buran of central Asia, and the pamperos of Argentina are some of the local names given to the persistent winds that often blow for days on end.

Savanna grasslands

Savanna grasslands, the world's tropical grasslands, lie closer to the equator. As a result, temperatures do not vary during the year as much as they do in temperate grasslands, and the weather is seldom very cold. In Cloncurry, Australia, for example, summer temperatures are usually about 86°F and winter temperatures about 60°F.

The particular feature of savanna grasslands is the seasonal rain. In winter, dry air sinks over the land, forming a block of high pressure that keeps out moist winds. This period is known as the dry season. Rain hardly ever falls, and plants gradually turn brown and their leaves die off. Summer brings the sun almost overhead, heating the air, which then rises and creates a zone of low pressure (see diagram). Wet rain-carrying winds from the equatorial regions surge in, and the rainy season arrives.

Below In tropical regions when the sun is overhead in summer a zone of low air pressure forms: this can cause rainfall. In winter months, high-pressure zones occur. These block rain-carrying winds, making winter the dry season.

Descending air is very dry. It creates a high-pressure zone.

High pressure

June— sun overhead

Tropic of Cancer

Low pressure

When the sun is overhead, the air heats and rises, causing a low-pressure zone. As air rises, it cools and forms moisture, which falls as rain.

Equator

High pressure

December— sun overhead

Tropic of Capricorn

The rainy season can come suddenly in heavy storms that bring torrents of rain and fill rivers that are dry for the rest of the year. The dry brown landscape quickly turns green as fresh shoots begin to sprout.

Savanna rainfall is not the same all over the world—it can vary from as little as 10 inches to as much as 60 inches. The farther grasslands lie from the equator, the drier the climate becomes and the shorter the rainy season. Rainfall also becomes more unreliable. In recent years, parts of Africa have had rainfall well below normal and even years of drought. The Sahel, a region of dry savanna lying just to the south of the Sahara Desert, is one area that has suffered badly as a result. Without rain, grasses have shriveled up, animals have died, and the people have faced starvation.

Storm clouds build up over the dry African savanna at the beginning of the rainy season.

Spread of grasslands

Many of the areas that are now grassland are thought to have once been covered by forests. A great change in the earth's climate took place between 65 million years ago (when dinosaurs became extinct) and 35 million years ago. It is thought that one of the main reasons for the climatic change was the creation of great mountain ranges such as the Himalayas, the Rockies, and the Andes. These formed new barriers to rain-carrying winds, making the climate inland much drier. Grasses first began to appear about 55 million years ago

and were ideally adapted to surviving in the drier climate. Without sufficient water, trees died off and grasses took their place.

Fire can spread grassland, particularly in savanna regions. In these tropical areas, when storms build up at the end of the dry season and before the rains arrive, lightning can strike in forests. The fires that are started destroy most trees and other large, slow-growing plants. But grasses soon grow back when the rains come again, spreading over the burned land, replacing forest with grassland.

Grazing wild animals have also played a part in encouraging grasses. The animals eat not only the grass but also young growing shrubs and trees. The shrubs and trees do not survive being nibbled down to the ground and are killed off, but grasses grow even more strongly because they thrive on being regularly grazed and cut down.

It is not just these natural conditions that have created grasslands. For thousands of years people have cut down trees for fuel and shelter and have also learned that by burning down forests they can increase the area of grazing for their domesticated animals.

Farmers in Africa burn down trees and shrubs, leaving cleared land where grasses can grow. This will give them more pasture for their cattle.

3. GRASSLAND SOILS AND PLANTS

A small farm on the Great Plains of Hungary in eastern Europe. The area was once natural grassland but it is now farmed. The dark soil is especially rich and crops grow well.

Soils

Not all soils are the same. If a soil is rich, most plants grow well. Rich soils are able to hold water better and are naturally more fertile. Much depends on the amount of humus in the soil. Humus is the crumbly remains of broken down and rotting vegetable and animal matter that becomes mixed in with soil particles. Healthy soil with plenty of humus has a good structure and is usually fertile.

Some temperate grasslands have especially rich soils. Large stretches of the eastern European steppes are covered by dark soils called chernozem soils. Water is trapped in the dense mat of grass roots near the surface, and rotting roots and other organic matter are broken down, helped by the action of worms and insects. A layer of dark, very fertile soil is the result. In North America, large areas of the prairies also have dark brown or black chernozem soils.

Soils in savanna regions generally tend to be poorer with less humus. Heat from the sun and moisture in wetter savanna areas rot the organic material very quickly, and it has little chance to pile up into thick layers. Many savanna soils have a red color that comes from iron in the ground combining with water.

Grasses and other plants are important in the process of enriching the soil and because they hold the soil together. If plants are removed, wind and water can blow or wash the soil away in a process called soil erosion. Once soil erosion begins, it is often difficult for vegetation to take hold again, and land can turn into desert.

Grasses

Grasses are among the most common plants in the world and are able to survive in severe climates. On the prairies and steppes, grasses suffer extremes of heat and cold whereas in savanna regions, drought is often their main enemy.

Grasses can grow with far less water and warmth than are needed by trees. They survive because a large part of the plant—a mass of roots—remains underground. In cold or dry weather, the stems and leaves of grasses can die off but the plant sends up new shoots when conditions improve. Because grasses are firmly fixed by roots and have tops that can bend, they are able to withstand violent winds.

Many grasses reproduce by sending out underground stems or long shoots from which a new plant grows. They can do this for years without producing seeds. If the leaves are withered by drought or fire, nibbled by grazing animals, or cut by humans, the grasses are able to grow again.

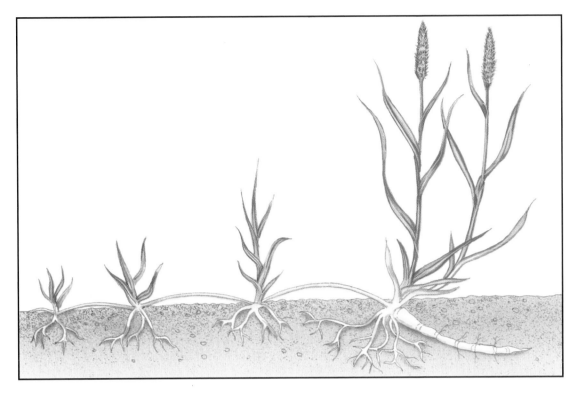

Left Most grasses have a mass of roots underground. If the top of the grass is cut, eaten, or destroyed by fire, a new plant can grow from the underground shoots.

Below Grazing wild animals, such as this zebra, can actually encourage grasses to grow more strongly.

Most grasses benefit from being cut, growing thicker and faster afterward, like a garden lawn that is frequently mowed. Some intertwine to form turf, a huge expanse of grass like a carpet, whereas others grow in coarse clumps known as tussocks.

Some grasses reproduce each year by letting their seeds scatter in the wind. The seeds are able to survive for a long time in the ground without rain and then start sprouting as soon as they are watered. Rains do not last long and the growing season is short, so these grass plants go through their life cycle producing flowers and seeds very quickly.

There are at least ten thousand different species of grass in the world. In areas of natural temperate grassland, grasses have deep roots, and they often grow as turf. Depending on the rainfall, grasses

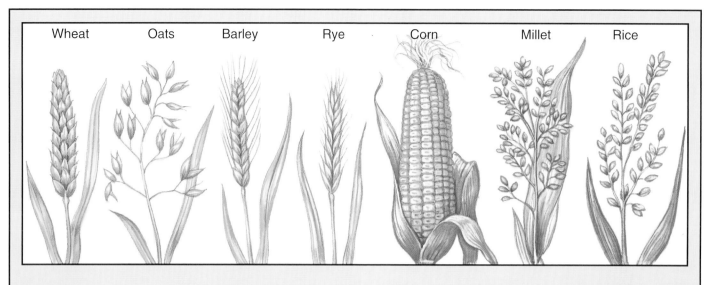

Wheat	Oats	Barley	Rye	Corn	Millet	Rice

Cereals

Thousands of years ago, humans discovered that some grass seeds, when dried and stored, provided nourishing food. These are now the cultivated grasses or cereals (shown above) that we depend on for food for ourselves and our animals. More than half the world's population relies on the grain of just three grasses—wheat, corn (also known as maize), and rice. These are used to make our daily food such as bread, pasta, mealies (a kind of cornmeal mush eaten in Africa), and rice.

may grow up to six feet high, or may grow just a few inches like the buffalo grass in the dry western prairies of North America. Most grasses have adapted to survive bitter cold.

Grasses in savanna regions vary a great deal in height, from short tufted grasses at the margins of deserts to tall elephant grasses that sometimes reach up to twelve feet in wetter savanna areas.

Red mounds of soil dot this grassland in Brazil. They are common in savanna areas and are caused by termites. Termites push the earth up above the ground and make their nests in the mounds.

These American temperate grasslands are filled with flowers during the summer months.

Other plants

Several other kinds of plants manage to grow in among the grass. In temperate grasslands, flowers such as the prairie crocus bloom in spring while later in the year clovers, poppies, and daisies appear. On some steppe grasslands, herbs nestle among the grasses and give off a strong scent if they are stepped on.

Most savanna regions have trees and shrubs that have adapted like grasses to the low rainfall. Many have deep roots to tap what water there is in the soil. Trees such as the baobab tree store water in their trunks, so that they can survive through the long dry season. With so little water around, baobabs are very slow growing and can live for hundreds of years. They also have tough bark to protect them from fire. In the dry season the baobab, like many other savanna trees, sheds its leaves.

A baobab tree in the Australian outback. It has a thick trunk that can store water, so that the tree survives through the dry season.

17

One especially common tree in Africa is the acacia. Each acacia produces a huge number of seeds every year (one variety releases as many as 20,000 seeds) to allow for the fact that few will survive. As protection against grazing animals, bushes and trees such as the acacia have developed thorns on their branches.

Soil creatures
Insects and small creatures such as earthworms play a very important part in breaking the soil up and turning it over, letting in air and water and putting nutrients back into the ground. In savanna grasslands, one of the main insects that turns soil over is the termite. Looking rather like an ant, the termite drags earth from under the surface to make tall nests above ground. Inside these mounds is a network of tunnels with chimneylike air vents that let in air and keep the nests cool.

Inset Acacia seeds form inside pods; when they are ripe, the seeds scatter over the savanna.

A small acacia tree, one of the most common trees in the African savanna. The long thorns protect the tree from most animals, but this hungry springbok is braving the spikes because the grass is very poor at the end of the dry season.

4. GRASSLAND WILDLIFE

Grasses provide food for all sorts of animals, and there was a time when grasslands all over the world teemed with wildlife. Many creatures have now disappeared, mainly because humans have hunted them down or turned grasslands into farmland, which has destroyed the natural habitat of the animals that lived there. Despite this, some natural grassland habitats remain, home to all kinds of mammals, insects, and birds. Savanna grasslands tend to have more animals than temperate grasslands have because they are relatively untouched by people.

Above Giraffes browse the tops of trees, their long necks allowing them to take the tender leaves that other animals cannot reach.

Large herbivores

A herbivore is an animal that eats grass or other plants. In the savanna grasslands of Africa, the largest herbivores are elephants, giraffes, and rhinos. Giraffes and elephants are browsers, feeding on trees and shrubs that grass-eating animals will not eat or cannot reach. Their main defense against predators is their size: few animals venture to attack such huge beasts. However, the most common of the large herbivores in both savanna and temperate grasslands are the grazers, grass-eating animals that include the various species of antelope, bison, and horse. They have all developed strong grinding teeth to deal with grasses that are tougher to eat than leaves.

Left Elephants by a waterhole in Botswana, Africa. Most wild animals in the savanna are now only found in areas of grassland that are wildlife reserves. Here they can live in their natural habitat undisturbed by people.

A herd of wildebeest cross a river in Africa. Like zebra and antelope, they live in large herds, grazing the savanna grass.

Grazers such as gazelle, eland, and impala are some of the species of antelope that roam across the African savanna. Like the wildebeest and the zebra, they live in large herds, which gives them some protection against predators. But their main defense when danger threatens is for the herd to close up together and run. Many of these animals migrate as the seasons change, traveling as much as 125 miles from one part of the savanna to another to find fresh grass and water.

Wild horses

Wild horses once roamed the grasslands of central Asia where they were domesticated about five thousand years ago. They were tamed and used by the Mongols, a race of nomadic herdsmen who not only became skilled at horseback riding but were also fierce warriors and conquerors. Today, the people of Mongolia still keep herds of horses. There used to be wild horses in North America as well, but they died out about ten thousand years ago, and it was not until Spanish settlers brought horses across the Atlantic in the sixteenth century that they again grazed the American grasslands.

A Mongolian girl on horseback herds her flock of sheep.

The prairies and steppes also once had herds of grazing animals such as antelope and horses, though not the very large animals that are found in Africa. Up until the nineteenth century, it is estimated that over 50 million American buffalo (bison) roamed the grassy prairies of North America. But when settlers spread west across America in the early 1800s, they began to hunt the animals down. By the end of the nineteenth century, less than one thousand buffalo were left, and they nearly died out altogether. Now buffalo are protected in reserves, and their numbers have grown.

Each region of grassland around the world has its own particular species of large herbivores: in Australia there are kangaroos and wallabies, on the Russian steppes the strange-looking saiga antelope, and in the high Andes Mountains of South America are llamas with their thick wool that keeps out the bitter cold.

Above The kangaroo is Australia's best-known grazing and browsing animal. It uses its powerful back legs for jumping, and its long tail helps it to keep its balance.

Large carnivores

The wildlife of grasslands depends on a food chain: there are animals that eat plants and other animals that prey on the plant eaters in turn. The meat eaters are known as carnivores. Most large carnivores are found in the savanna grasslands of Africa and include lions, cheetahs, leopards, and hyenas. Their prey are mostly wildebeest, gazelles, and zebras. On the North American prairies coyotes are the main large carnivores; wolves, once abundant, have been hunted almost to extinction.

Right In the African savanna, carnivores (meat eaters) like these lions feed mainly on grazing animals such as zebra and antelope.

Lions tend to live in groups of up to 30 animals known as prides, and they often hunt together. Predators like these all depend on surprise and speed to catch their prey. The cheetah is the fastest animal on land over a short distance: it can sprint at 65 m.p.h. but it cannot keep this speed up for long. Hyenas attack grazing animals but they are also scavengers, crowding around, waiting to finish off the remains of a kill that another predator has already made.

Above A cheetah brings down a gazelle. Cheetahs are among the fastest predators in the world.

Small mammals

Natural grasslands are also home to many small mammals. Most common are the various species of rabbits, gerbils, hamsters, mice, and ground squirrels, but there are slightly larger creatures, such as cavies in the South American pampas and wombats in Australia. Most live off grass, seeds, and leaves, although the anteater that lives in the pampas has a long nose and tongue specially designed to get at nests of termites. Small mammals also become food themselves for predators like foxes, weasels, and birds of prey.

Left This North American prairie dog is one of the many small mammals living in grasslands all over the world. They make their homes in burrows beneath the ground.

Saiga
The saiga is an antelope living in herds on the Russian steppes. It is about the size of a small goat, pale brown in color, and with bulging eyes, a huge snout, hooves, and horns. At the beginning of the twentieth century only a few hundred saigas remained because they had been hunted almost to extinction, mostly for their horns, which were greatly valued in China for medicine. In 1919 the Russian government passed laws to protect the saiga. They were saved just in time and, from only a few hundred remaining animals, they now number close to a million.

To avoid the heat and cold as well as to escape predators, many small mammals burrow into the ground. Some also use their burrows to store food and hibernate there in winter. The North American prairie dog (a kind of ground squirrel with a barking call) lives in colonies of thousands, making a vast network of tunnels below the surface of the ground. Farmers often consider small mammals like these to be pests because they not only eat grass and crops, but also make hundreds of holes in the fields.

Insects and birds
In among the grass lies another layer of life: insects. Hordes of flies, beetles, ants, grasshoppers, and caterpillars feed off various parts of the grass and other plants. The amount they eat can be far more than the amount the herds of large grazing herbivores consume.

A weaverbird hanging from the bottom of its nest, which has been woven from savanna grasses. Weavers and other birds eat the insects and grass seeds that are in plentiful supply.

23

Many insects also perform a very useful function in the soil. They eat the droppings left by large mammals or the rotting remains of dead creatures and turn these back into nutrients that in turn feed the plants.

The plentiful seeds and insects provide food for many species of birds, especially in the savanna. Some, like weaverbirds in Africa and budgerigars in Australia, live in huge flocks. There are also birds such as swallows, which migrate by the thousands every year to Africa to escape the cold winters of northern Europe and Asia. With so many small animals, grasslands make an especially good habitat for birds of prey such as eagles, hawks, falcons, and the scavenging vultures that live in the savanna. Like hyenas, vultures usually wait for other animals to make a kill before moving in to pick at the remains.

The savanna is home to the largest bird in the world, the ostrich. Often measuring up to six feet tall, ostriches cannot fly, so they make their nests on the ground. The hen bird may lay 20 or more enormous eggs. Ostriches graze on the savanna grasses, and when they sense danger they can run at great speed.

Ostriches are often seen strutting across savanna grasslands in parts of Africa. This pair have some very young chicks with them.

5. THE CHANGING LIFESTYLES OF GRASSLAND PEOPLES

Hunters and herders

Grasslands have never been densely populated because of the lack of a regular water supply. For the same reason, nearly all the people who lived there in the past led wandering, nomadic lives, moving around in a continual search for food and water and for fresh grass if they kept animals.

Until Europeans arrived in Australia at the end of the eighteenth century, the native people, the Aborigines, lived in areas of scrub savanna as hunters and gatherers. With bows and arrows they skillfully stalked wild animals and collected the fruit and roots of plants for food. They knew where to find water and food even in the driest areas on the edge of deserts. However, if they were to find enough to survive, they always had to be on the move and needed to be able to roam freely across large areas of land. One reason this way of life has almost died out today is that so much land has been fenced and used for farming.

Native Americans were once also grassland hunters. They depended on the thousands of buffalo that

The Aborigines, the native people of Australia, were once nomadic hunters and gatherers. The arrival of the British in 1788 caused a tremendous change in the Aborigines' way of life.

In the grasslands of Asia, Mongolian herders live in tents called yurts.

Traditional herders

The majority of people in Mongolia are farmers who do not grow crops but raise animals on the grasslands that cover huge parts of the country. Two-thirds of the animals are sheep but they also keep goats, cattle, horses, and camels. Mongolia has a very dry climate (about eight inches of rain a year), and temperatures average between -20°F and 60°F. Many herders still follow the traditional way of life, moving from pasture to pasture with their animals, living in thickly lined tents (called yurts or gers) that keep cool in summer but warm in the bitterly cold winters.

roamed wild across the prairies for food. They used the animals' skins and bones to make clothes, tents, weapons, and jewelry. Not surprisingly, the buffalo was greatly honored by many tribes. Until horses were brought to the Americas by Europeans, Native Americans used to hunt on foot with spears and arrows.

In central Asia herding animals is a way of life that has been followed for centuries. In drier grasslands that have not been cultivated, people still live in this way. From their domesticated flocks of animals, such as sheep, goats, horses, and even camels, people get milk, butter, and cheese as well as meat and skins that they trade for other goods. Animal dung is used as fuel because of the lack of wood in these treeless spaces. The Mongolian and Kazakh people, for example, live in tents, because they have to be able to move their homes easily when they need to find fresh pasture for their animals.

This picture, painted in 1835, shows Native Americans on horseback hunting buffalo on the prairies. Their traditional way of life was destroyed when Europeans killed off millions of buffalo to clear the prairies for their farms.

New arrivals

During the nineteenth century great changes took place in many of the world's temperate grasslands. People were leaving Europe in huge numbers to escape poverty and war and in the hope of making a better life in a new land. Some who moved to North America went west toward the prairies; in South America they settled in grassland areas such as the pampas of Argentina. Others traveled eastward from Europe, spreading across the steppes of Russia.

Laying the railroad in the United States in the 1870s. The railroads made it easier to travel across the vast plains and speeded up the settlement of the prairies.

Great herds of buffalo had for centuries grazed the prairie grasslands, but few escaped the slaughter by European settlers. These animals, grazing quietly in the safety of Yellowstone National Park in Wyoming, are among the few remaining.

Until the arrival of the European settlers, the prairies of North America were open grassland with tall grasses and flowers, grazed by the buffalo that were hunted by tribes of Native Americans. Early European explorers thought that the endless expanses of waterless and treeless grassland were of little use. The French first realized that the plains might be useful for grazing domesticated animals and gave the region its name, *prairie*, a French word meaning meadow.

As waves of settlers arrived in the United States and Canada in the nineteenth century, the Native Americans were forced off the plains, often after fighting bloody battles to try to keep their lands. Great massacres of the buffalo began, to clear the wild animals off the land and to create grazing land for the newcomers to farm their cattle.

European settlers, known as sodbusters, began to plow up the grasslands of North America and plant crops.

Technology and inventions

The building of railroads across the North American prairies in the 1860s spurred on settlement and farming. Instead of driving cattle across vast distances on foot, live animals could now be transported easily by rail to market. North America's growing population demanded more and more

meat, and cattle ranching became big business. Settlers poured onto the prairies where land was still cheap. Large cities like Chicago, Illinois, grew up around railroad stations.

Gradually farmers realized how rich the prairie soils were and that they could be cultivated as well as used for grazing cattle. People started to plow the land and grow cereals, but it was difficult to work the vast fields by hand. When suitable machinery that could do the plowing, sowing, and harvesting became available in the 1930s, the prairies were turned over to large-scale crop farming.

There was new technology of other kinds that gradually made it easier for people to settle in and farm temperate grasslands such as the prairies. Rivers and lakes were dammed and pipelines brought water to homes and farms. In more remote areas, underground sources of water were tapped by using wind-powered and other mechanical pumps.

Because meat spoils very quickly unless it is preserved, freezing and canning were very important inventions for cattle farmers, especially in places such as the pampas of Argentina, far away from world markets. In the late nineteenth century the first shipments of frozen beef reached Europe from South America.

A cowgirl rounds up horses. Some farmers, in the drier western prairies, still use horses for herding cattle in much the same way as the first cattle ranchers.

Most of the prairie grasslands have now been turned over to large-scale crop farming.

New technology and inventions speeded up the development of grasslands all over the world, not just in the Americas. The Trans-Siberian railroad was begun in 1891 and completed in 1905, crossing the Russian steppes and bringing land within easy reach of the stream of settlers that followed the building of the railroad. During the 1950s, the use of large-scale machinery enabled vast areas of the steppes to be turned over to crops such as wheat and barley.

The demand for meat and grain from the industrial regions of the world grew, encouraging more intensive farming of grasslands. By the mid-twentieth century, nearly all the once-wild prairies had been plowed and cultivated, and either crops or herds of cattle and sheep covered large stretches of temperate grasslands all over the world.

Pressure on the savanna

The grasslands of Australia were also affected by the spread of European settlers during the nineteenth century. The new arrivals took and fenced land for their sheep and cattle ranches, and, as in the prairies, the traditional way of life for the native peoples was changed forever.

Europeans settled in parts of Africa as well, driving the original inhabitants off their land. But large-scale ranching in the savanna never took off quite as it did in North America or Australia, partly because tropical diseases attacked the plants and animals that Europeans introduced—for example, sleeping

Above Australian Aborigines round up cattle on a large farm in the outback.

Above A cattle market in Mali, Africa. The way of life in the African savanna has changed less than in other grassland areas of the world.

sickness caused by the tsetse fly was, and still is, a major problem for both cattle and people.

Although Africans have for centuries used the grasslands as either hunter-gatherers, pastoralists (another name for people who herd animals), or small-scale farmers, there was always a balance between people, animals, and land. There were not too many people and they used the savanna in a way that did not destroy the natural habitat. As a result, Africa is one of the few places in the world where there are large areas of natural, relatively unspoiled grassland.

However, since the mid-twentieth century the population has grown rapidly and most of the people still depend on farming to survive. To produce enough food to feed so many people, more and more land is being used for crops or for grazing domesticated animals, and the natural savanna grasslands and their wildlife are fast disappearing.

Herders to farmers
Some African tribes such as the Masai and Fulani are, by tradition, herders. The people used to migrate with their herds of cattle across the savannas each season but many can no longer find enough grazing land. Some have had to settle down and turn to planting and farming crops, often on infertile land. Many of these tribespeople know little about growing crops. If they fail to make a living off the land, they have no choice but to move to the towns where they will almost certainly have to live in slum conditions and try to survive on the streets. Their age-old way of life is changing quickly.

A Masai boy tends his family's goats in Kenya, Africa.

6. FARMING GRASSLANDS

Seas of grain

Once carpeted in grass, the prairies of North America and the steppes of Russia and Ukraine have become some of the greatest wheat-producing regions of the world.

Wheat is the main cereal, but farmers also grow corn, barley, rye, and oats. Although these cereals are related to grasses, they are very different from their wild ancestors. They have been scientifically bred to produce a heavy crop, to ripen quickly, and to withstand pests and diseases.

Wheat is mostly used to make flour (used in bread, pasta, etc.). Corn, with its big yellow grains that cluster together in a cob, is grown for other reasons: it can be used to make breakfast cereals, popcorn, certain chemicals and glues, and cooking oil. One of the largest corn producing areas is the United States's corn belt, in the eastern prairies where the rainfall is slightly higher than in the west. Much of the corn from this area is used as cattle fodder.

Both wheat and corn grow best in warm summers, so where summers are cooler, such as in the northern steppes of Russia, barley, oats, and rye are often grown. These cereals can be used for animal feed and for making alcoholic drinks like beer and whiskey.

Several combine harvesters are needed to cut the crops of wheat and barley on these enormous fields.

Crops are sprayed with chemicals to keep them free from pests and diseases. The only practical way to cover the vast fields in places like the prairies is to use airplanes called crop dusters.

In the southern prairies, wheat is planted in the autumn, but in the northern plains of the United States and Canada, where the winters are very cold, it is sown in the spring. The crop has to ripen quickly during the summer to be ready for harvesting in the autumn. Fertilizers are spread on the fields to help the wheat to grow, and pesticides are sprayed on the plants to keep them free from disease and pests. Often this is done from airplanes because the fields are so vast.

Modern wheat farming is big business, using expensive and highly advanced machinery on farms that cover hundreds of acres. To harvest the wheat at exactly the right time, farmers work nonstop, sometimes using combine harvesters with headlights to work through the night. The harvested grain is stored in huge towers called elevators or silos. Later it is loaded onto trains to be taken to mills where it is ground into flour or to ports for shipping abroad.

Because rainfall is often low and uncertain in grassland areas, crops sometimes need to be irrigated. These huge pipes are used to carry water for spraying onto the surrounding crops.

Other crops, such as sugar beets and soybeans, are grown as well. Some farmers rotate their crops, perhaps growing wheat or corn one year and then plants such as alfalfa, clovers, or ryegrass another year. Crops such as wheat and corn quickly use up the natural nutrients in the soil, and these other plants, which are used as animal fodder, help to put the nutrients back.

A water supply is important because rainfall is not reliable and in times of drought crops may have to be watered artificially. Sometimes fields are irrigated using reservoir water from a dammed river, which is often piped or channeled from long distances away. Often the only source of water is an aquifer, which is a natural reservoir of water stored underground in rocks. A well or pipe is sunk into the ground and water is pumped out and sprayed onto the fields, using pipes and sprinklers that can cover hundreds of acres.

Herds of animals

Not all temperate grasslands have been plowed up for crops; many are still used for raising cattle or sheep. In the United States, cattle are mainly kept on the drier fringes of the grasslands, where there is too little rain to grow crops. It is important to have a water supply for the animals, and wind pumps are often used to bring water up from underground to a trough where the animals can drink.

Domesticated cows have been selectively bred from the lean wild cattle that were their ancestors to produce animals that give good meat despite the relatively dry conditions of temperate grasslands. The most common breeds of cattle are Herefords, Angus, and Charbray, which are raised for their meat and not for their milk. Milk cows, such as Holsteins and Jerseys, need richer grass such as that in the much wetter cultivated pastures of northern Europe.

Cattle farming is big business in parts of Australia. Here the animals are penned up ready for sale at a market.

Lassoing a calf on a cattle ranch in the llanos grasslands of Venezuela, South America

Ranches usually cover vast areas, but the number of animals per acre in drier grasslands is often low to prevent overgrazing of the poor grass. Fences that stretch hundreds of miles need to be checked regularly, and animals have to be brought in to be branded or tagged with their owner's mark in case they should stray. There are still parts of the world where cattle are herded by cowboys and cowgirls on horseback, although many farmers now use vehicles.

The most important animals raised in the steppes of central Asia and in Australia and New Zealand are sheep. Australia is the biggest sheep-producing nation in the world. The animals, which can graze poorer grassland than cattle, are reared both for wool and meat. There are many different breeds of sheep; some are bred to survive cold winters, while others are bred specially for their fine wool or meat.

A sheep farm in New South Wales, Australia

African savanna farming

In parts of the African savanna, farming has changed little over hundreds of years. Many people keep small herds of cattle or goats, continually moving them around in the search for fresh grass. Others grow crops, cultivating new patches of savanna every few years because the soil becomes worn out and needs to rest to recover its fertility (this is called crop rotation, or shifting cultivation). These kinds of farming need large areas of land so that there is always some fresh land available. It is also subsistence farming, that is, farming to produce food for families to live on. There are no large machines to do the hard work, and the land has to be cultivated by hand.

Millet and sorghum are two of the most important cereals in savanna regions. These cereals are able to withstand heat and drought. Maize is cultivated where there is more water, and people grow other crops such as

Above Maize is an important crop grown by farmers in Africa. The people do not have machinery and are putting fertilizer on the fields by hand.

Left African villagers pound maize to make flour.

Hardy cereals
Both millet and sorghum are useful cereals for farmers in dry savanna regions because they need relatively little water. They are planted as the wet season begins and when fully grown the plants can reach up to 12 feet in height. The grain is usually crushed and boiled and made into a kind of cereal or into a flatbread. The stalks are sometimes used as fodder for animals.

yams, cassava, and vegetables as well.

Farmers also plant cash crops. These crops are not used by the people who cultivate them but are grown to be sold. They include tobacco, cotton, peanuts, sisal, sugarcane, and coffee. Some of these will grow quickly during the relatively short rainy season. Others need more water, and they are planted on the best watered land, often near a river. Poorer farmers can seldom afford this well-irrigated land and it is usually owned by wealthy farmers with large landholdings.

In some areas, especially in southern Africa where the savanna merges into temperate grasslands, landowners have fenced the land and use it to raise beef cattle on ranches. Many farmers use the white Brahman cattle that were developed from breeds from India because they can resist heat and tropical diseases better than other breeds. These are often crossbred with local breeds and European breeds to produce animals that cope well in the difficult savanna conditions.

A herd of cattle in Kenya. The animals are often moved around the savanna grasslands to provide fresh grass.

Pest of the savanna
Locusts are a kind of grasshopper. Rather like nomads, they travel around looking for food, sometimes covering great distances at great speed. They live in the dry grasslands and desert margins of Africa and Asia and appear in great swarms after heavy rainfalls. As many as 40 billion locusts can gather in one swarm, and within days they can destroy fields of crops and ruin harvests. Locust swarms are often controlled by spraying with pesticides but often not all of them are killed.

7. USE AND MISUSE OF GRASSLANDS

When the grasslands of the prairies were first plowed up and crops planted, several years with good rain followed, bringing rich harvests. The pioneering settlers, called sodbusters, dug up the land and sowed crops year after year, using north European farming methods that depended on plentiful rain. But then, during the 1930s, came drought.

Dust Bowl

Between 1917 and 1939, there were 17 years when rainfall was well below normal. Plants withered and died. The hot sun dried out the soil, leaving no moisture to hold it together. Unprotected by plants, the topsoil was whipped up by the wind and carried away in tremendous dust clouds across the prairies. Dust storms swept thousands of miles across the United States. The western prairies ended up like a desert and became known as the Dust Bowl.

By the 1930s, when this photograph was taken, large areas of the prairies had been plowed. There were several years of drought and plants died. The dry topsoil swept like dust across the plains.

With the failure of their crops, huge numbers of farmers and their families were left penniless and starving. Many left the land to seek work in towns and cities. There were also serious long-term effects on the land because the fertile topsoil had been blown away. Although the land has now recovered, it has taken many years of careful management and control.

Overuse

The unreliable rainfall in many grassland regions of the world and the pressure on the land from growing populations is producing similar "dust bowl" problems in other parts of the world. In particular, in Africa since the 1960s the numbers of people and domesticated animals have doubled. As a result, many areas have been overgrazed, destroying the plant cover and leading to erosion of the soil.

In parts of Africa, like the Sahel, there are too many animals kept on poor grassland. Rainfall is low and the grass is soon eaten, so that it becomes more and more difficult for the people to find enough grazing for their sheep, goats, and cattle.

The poor grassland of the Sahel (see chapter 2) in Africa is one example of this problem. It is traditionally grazed by cattle whose herders follow the rains north during the rainy season and retreat to greener land in the south during the dry season. Now, large areas of the wetter southern savanna have been turned over to crops, and the nomadic herders have lost their dry season grazing land. They have to rely on the northern drier parts of the Sahel all through the year for food for their animals. Too many animals on such poor land has stripped it of vegetation, and the low and irregular rainfall means that the soil never has a chance to recover. Large parts of the Sahel are turning into a wasteland.

Greedy goats

Many of the less well-off farmers in Africa keep herds of goats, which provide them with milk and meat. Goats munch their way through any greenery they can find, even woody plants and thorny shrubs, as well as grass and leaves. Although this makes them ideal animals for people who live on the poorest lands, they can cause great harm to the land. When they are left to eat whatever they like, goats have ended up stripping many places bare of plants, leaving a wasteland.

Overuse of the land has other effects. The prairies of North America were once fertile, but the continual use of the land for crops has taken nutrients out of the soil. In Africa, the pressure to grow more food means that land that was once left every few years to recover is now used all the time. As the soil loses its natural fertility, harvests become poorer and people end up with less to eat.

Looking after farmed grasslands

Many lessons were learned from the dust bowl disaster in the prairies, and so far the tragedy has not been repeated. Cultivating grasslands is not just a matter of plowing up the land and sowing any crop; careful farming management is needed to prevent problems.

Plowing very dry land can lead to erosion; so can plowing land in lines up and down slopes, which makes channels for water to pour down and wash away soil. One method to help prevent this problem is contour plowing, where land is plowed across the slope of the land. In other parts of the world, low stone walls are sometimes built to stop water from running off and taking soil with it. In the battle against wind erosion, farmers plow at right angles to the direction of the wind and then plant wheat and grass in alternate strips.

Harvesting alfalfa, which is a crop used for animal fodder. The farmer is working across the slope of the land so that the lines he leaves do not run downhill. When rain falls it is less likely to flow down the slope and wash the soil away.

Prairie farmers learned to leave wheat stubble and straw on the fields after harvesting, both to return some vegetable matter to the soil and to ensure that the soil is not left bare and open to the wind. Tree planting has also proved a success in many parts of the world, providing shelter from winds that not only erode the soil but also flatten crops. In Africa, native trees such as the acacia are planted because they are more able to cope with drought.

As soils lose their fertility due to more intensive farming methods, chemical fertilizers have to be added to the land every year to help crops grow. But, as well as being expensive, these chemicals can be harmful if they find their way through the soil into drinking water supplies. There is pressure on farmers in the United States not to use so many chemicals and to use animal manure instead, which is a natural fertilizer. Other chemicals used to control pests that attack cereal crops can also be dangerous and upset the balance of nature. One solution that may help is the development of new disease-resistant varieties of wheat and other cereals.

A more natural way of restoring the soil's fertility is by crop rotation, planting the fields every other year with deep-rooted plants such as alfalfa and clover that put nutrients back into the soil. Mixed farming, such as growing crops and raising cattle, also helps bring a balance to the land and means that the farmer does not have to rely on just one source of income in case a crop fails.

African farmers in the Sahel are shown how to build low stone walls around their crops to try to stop water from running off and causing soil erosion.

Dealing with these problems in the African savanna is more difficult because farmers have less money to spend on expensive aids such as chemical fertilizers. There is also less understanding of how intensive farming can harm the land. Cheaper and simpler farming methods have to be used (for example, building low stone walls to prevent runoff), and using local knowledge and techniques often produces the most successful results. Even so, producing enough from the land to feed the rapidly increasing population is not an easy problem to solve.

Apart from the need for care with crop farming on grasslands, solutions have to be found to the problem of overgrazing. The best answer is to allow the land to rest and give it time to recover. This is done in some places by controlling grazing land with fencing. However, in poor areas like the Sahel, where population pressures are so great and the rainfall is so low, it may be too late to save the dry grassland from turning into desert and the herders who once depended on it from losing their traditional means of farming.

A patchwork of fields in North America. One year wheat may be planted in a field and the next the crop may be alfalfa or clover. By changing the crops like this the farmers can help maintain the natural balance of chemicals in the soil.

Disappearing wildlife

One of the problems in grassland areas is the conflict between humans and wildlife. Some argue that natural areas should be kept so that animals can continue to live in their natural habitat. Otherwise the only wildlife left in the world will be in zoos. But the bigger grassland animals need a large area of grazing, and people are not always willing to spare land that could be used to produce food.

As we have seen, the prairie settlers thought the land could be better used to raise cattle or grow crops than to leave it to wild animals. As a result, there is now virtually no natural grassland left in North America. As well as killing off large animals such as the buffalo, people have also destroyed the grasses and plants that form the natural habitat for small grassland creatures. This has happened in much of the world's temperate grassland.

The saiga antelope of the Russian steppes (see page 23) is an interesting example of how a native grassland animal can remain in its natural habitat and yet be useful. Luckily, before it became extinct, people came to realize that it was a useful source of meat and that leather could be made from its hide. Now saiga are farmed and every year about 250,000 are killed.

This kind of solution could be used in many parts of the world if people were prepared to eat the meat of other animals instead of cattle. Buffalo could be bred in the prairies, and there are many varieties of antelope in Africa that produce good meat. Raising cattle is particularly difficult in Africa, with the problems of disease and the lack of good all-year grazing, and it would be more sensible to use native animals that are adapted to this habitat. This would help preserve not only these large animals but also the natural grasslands themselves and the many other smaller animals that live there.

Few areas of natural grassland remain in North America. In this snow-covered wildlife reserve, buffalo are protected and can live in their natural habitat.

Wildlife reserves

Most natural grassland wildlife only survives where it is protected in reserves. In the prairies there are some reserves where buffalo graze natural prairie grasses and wildflowers still bloom, but these areas are not large enough for herds of migrating antelope, and bears and wolves no longer exist. However, because the world now depends on the prairies and steppes for so much of its wheat, and because so much wildlife has been destroyed, it would be impossible to return these grasslands to their natural state and bring back the wildlife.

Only the African savanna has large protected areas of unspoiled grasslands where wildlife lives relatively undisturbed. But even here animals are under threat. Hunters shoot rhinoceroses for their horns (believed to be a valuable medicine in some parts of the world); elephants are killed for their ivory tusks; spotted cats like leopards are hunted for their beautiful coats. Some animals are shot for their meat. Although game wardens patrol against hunters, it is difficult to keep watch on such vast wild areas.

However, the greatest threat to the African grasslands is that people want the land for farming. In many parts of the continent people are desperately poor, their numbers are growing, and they need the land to grow enough to eat. It is understandable that these people see no reason why the animals should be protected when they are starving.

Tourism may be the best solution to preserving Africa's last savanna grasslands and their unique wildlife. Many African countries are encouraging tourists to visit their game reserves to see wild animals in their natural habitat. If tourist money goes to the local people and provides them with jobs, the savannas may be seen as having more value left as they are than if they are turned into farmland. This may provide a balance between the need for people to make a living and preserving the world's last natural grasslands.

Above Tourists watching lions. It is hoped that the money the tourists pay to visit African game parks and see the wildlife may encourage local people not to turn these unspoiled grasslands into farmland.

A black rhino and its calf in an African game park. Rhinos are among the most endangered of all African animals.

Glossary

Altitude   The height of a place above sea level.

Bison   A large oxlike herbivore with a shaggy coat. Herds of bison, or American buffalo, used to graze the prairies in North America.

Browsers   Animals that feed off leaves from bushes and trees.

Cereals   Members of the grass family that are grown for food.

Chernozem soils   Dark, fertile soils found in temperate grasslands.

Cultivated   Used for growing crops.

Domesticated animals   Tamed animals used by people for food, transportation, etc.

Equatorial   Relating to the regions near the equator, the imaginary line that circles the middle of the earth.

Extinct   No longer existent. Describes a species that has died out completely.

Fertility   The level of plant nutrients in the soil.

Fertilizer   A substance added to farmland to make it better for growing crops.

Fodder   Coarse food fed to animals such as cows, sheep, and horses.

Habitat   A place that provides a natural home for plants and animals.

Hibernate   To spend the winter in a state similar to a deep sleep. Several kinds of animals hibernate to survive harsh winter weather.

Intensive farming   Farming the land to its fullest capacity; it usually means that fertilizers and irrigation are needed to keep the soil productive.

Irrigated   Watered by artificial methods.

Latitude   The distance measured in degrees of a place north or south of the equator.

Mammals   Warm-blooded animals that feed their young with milk from the mothers' bodies.

Migrate   To move from place to place, often at different times of year.

Nomadic   Describing people who regularly move from place to place to find food and water either for themselves or for their animals.

Nutrients   Substances in the soil that feed plants.

Overgrazing   Feeding so many animals on grass that there is no longer enough left to support them.

Pesticides   Chemicals used to kill insect pests.

Predators   Animals that hunt and kill other animals.

Prey   Animals that are caught and eaten by other animals.

Rain shadow   An area where rainfall is low because mountains block rain-carrying winds.

Reserves   Special areas set aside where wildlife and plants are protected.

Reservoir   A place where water is collected and stored.

Scavengers   Animals that feed on dead animals that they have not themselves hunted or killed.

Scrub   Bushes and stunted trees.

Seasonal   Happening only at certain times, or seasons, of the year.

Southern Hemisphere   The half of the earth that lies south of the equator.

Technology   The application of science for practical uses.

Temperate region   A region with a mild or moderate climate.

Tropical   Relating to the tropics—the area around the middle of the earth between the tropics of Cancer and Capricorn, north and south of the equator.

FURTHER READING

Amsel, Sheri. *Grasslands.* Habitats of the World. Milwaukee: Raintree Steck-Vaughn, 1992.

Dixon, Dougal. *The Changing Earth.* Young Geographer. New York: Thomson Learning, 1993.

Fleisher, Paul. *Ecology A to Z.* New York: Dillon Press, 1994.

Flint, David. *The Prairies and Their People.* People and Places. New York: Thomson Learning, 1994.

Khanduri, Kamini. *Grassland Wildlife.* World Wildlife. Tulsa: EDC Publishing, 1994.

Nile, Richard. *Australian Aborigines.* Threatened Cultures. Milwaukee: Raintree Steck-Vaughn, 1992.

Sayre, April Pulley. *Grassland.* Exploring Earth's Biomes. New York: Twenty-First Century Books, 1994.

Sita, Lisa. *Peoples of the Great Plains.* Celebrating America. New York: Thomson Learning, 1996.

Taylor, Dave. *Endangered Grassland Animals.* Endangered Animals. New York: Crabtree Publishing, 1992.

Zipko, Stephen J. *Toxic Threat: How Hazardous Substances Poison Our Lives.* New York: Julian Messner, 1990.

FURTHER INFORMATION

For further information about animals and their habitats that are under threat, contact the following environmental organizations:

Center for Environmental Education, Center for Marine Conservation, 1725 De Sales Street NW, Suite 500, Washington, DC 20036

Friends of the Earth (U.S.A.), 218 D Street SE, Washington, DC 20003

Greenpeace U.S.A., 1436 U Street NW, Washington, DC 20009

The National Wildflower Research Center, 2600 FM 973 North, Austin, TX 78725

The Nature Conservancy, 1815 North Lynn Street, Arlington, VA 22209

World Wildlife Fund, 1250 24th Street NW, Washington DC 20037

These organizations all campaign to protect wildlife and habitats throughout the world.

Picture acknowledgments
Eye Ubiquitous /L. Fordyce 8, /T. Page 9, /L. Fordyce *title page*, 14(lower), 22(lower), 28(top), 31; Peter Newark 27(top), 28(lower), 39; Photri /C.W. Biedel 17(top), 27(lower); Still Pictures *cover*, /J. Nelson 10, /P. Harrison 12, /M. Laboureur 19(lower), /M. & C. Denis-Huot 20(top), /S. Perm *contents page*, 20(lower), /Klein Hubert 21(top), /A. & C. Denis-Huot 21(lower), /F. Polking 22(top), /H. Daniel 23(top), /B. van den Brink 23(lower), /N. Ostergaard 25, /P. Harrison 32, / V. Dedet 34(lower), /B. Pambour 36(top), /F. Hazelhoff 38(top), /F. Lemmens 38(lower), /M. Edwards 40(top) 42; /D. Drain 44; Tony Stone Worldwide /R. Lynn 4–5, /N. Parfitt 11, /R. Smith 14(top), /S. & N. Geary 17(lower), /P. Kenward 24, /T.Davis 29, /G. Irving 30, /P. Tweedie 31(top), M. Kezar 33, /F. Prenzel 35, /R. Smith 36(lower), /I. Murphy 37(both), /A. Sacks 41, /D.R. Frazier 43, 45 (top), /N. Parfitt 45 (lower); Julia Waterlow 5, 15, 16, 18(both), 19(top), 20(top), 26, 40(lower); Wayland Picture Library/ D. Cumming 13; Zefa 6, /F. Damm 7, 34(top). The artwork on pages 4, 9, 10, 15, and 16 is by Peter Bull Design.

INDEX

Numbers in **bold** refer to photographs